The Expo Boutique Hotel
世博精品酒店

第二版

李壮 主编

中国林业出版社
China Forestry Publishing House

图书在版编目（CIP）数据

世博精品酒店 / 李壮主编 . -- 2 版 . -- 北京 : 中国林业出版社 , 2017.9

ISBN 978-7-5038-9256-1

Ⅰ . ①世… Ⅱ . ①李… Ⅲ . ①饭店－室内装饰设计－世界－图集 Ⅳ . ① TU247.4-64

中国版本图书馆 CIP 数据核字 (2017) 第 207231 号

中国林业出版社 · 建筑分社

策　　划：纪　亮
责任编辑：纪　亮　　王思源
封面设计：吴　璠

出版：中国林业出版社（100009 北京西城区德内大街刘海胡同 7 号）
网站：lycb.forestry.gov.cn
印刷：北京利丰雅高长城印刷有限公司
发行：中国林业出版社
电话：（010）8314 3518
版次：2017 年 9 月第 2 版
印次：2017 年 9 月第 1 次
开本：1/16
印张：20
字数：150 千字
定价：280.00 元

CONTENTS

上海柏悦酒店	Park Hyatt Shanghai	4
上海博雅酒店	Parkyard Hotel Shanghai	32
上海浦西洲际酒店	Inter Continental Shanghai Puxi	62
上海博地精品酒店	Bodi Boutique Hotel Shanghai	96
上海朗廷扬子精品酒店	The Langham Yangtze Shanghai	116
上海世茂皇家艾美酒店	Le Royal Meridien Shanghai	144
上海联艺·凯文公寓	Gallery Suites Shanghai	174
88新天地酒店	88Xintiandi Hotel	184
北京盘古七星酒店	Pangu 7Star Hotel Beijing	204
北京王府井希尔顿酒店	Hilton Wangfujing Beijing	238
上海半岛酒店	The Peninsula Shanghai	280
深圳大梅沙京基海湾酒店	Kingkey Palace Hotel Shenzhen	304

PARK HYATT SHANGHAI

上海柏悦酒店

地址：
上海市浦东新区世纪大道100号

LOCATION:
100 Century Avenue New Pudong, Shanghai

开业时间：
2009年

OPENING TIME:
2009

设计单位：
季裕堂(Tony Chi)

DESIGN UNITS:
Ji Yu Tang(Tony Chi)

面积：
26800平方米

AREA:
26800m²

上海柏悦酒店位于有(垂直型综合城区)之称的上海环球金融中心79至93楼,所处位置可谓是站在了巨人的肩上,将浦江两岸风光一览无余。酒店被定位为摩登的中国式精品住宅酒店,处处体现着它"低调的奢华"。

设计师不光考虑到酒店的内部设计,还考虑到客人进住酒店过程中的享受,因此一片长势茂盛的竹林立于此地,引人前行,前往心目中的目的地。

酒店的整体设计以灰色、白色、咖啡色为基调,以求低调与节约的华丽感,色系效果素雅和谐,并将富有创造力的艺术饰品贯穿其中,营造出一种中国水墨画的写意。

设计的经典在于整个空间设计没有典型的中式符号,却能表达足够现代的中式生活格局,完全颠覆了中国古典风格,让它以现代感十足的面貌展现在我们面前。

如果以居家的感觉来重新审视,底层的设计刚好是一段荡涤心灵回家的路,而每两个房间共享的一个空间,则是一个小小的天井,在套房的入口处所设计的小块草坪,有入户花园的感觉,再加上公共设施里的太极园,整个动向的分布就像是由大门进入,行经大厅与房间,最后终点于庭院。因为400 米之上的高度, 上海Park Hyatt 拥有世界最高的游泳池。万家灯火之上的那一方蔚蓝,就像漂浮于城市上空的一泓海水,这种意境在上海Park Hyatt 无疑更加生动,蓝色的水体静卧于室内的白与室外的黑之间,像一件纯粹的艺术品将空无的力量直抵人心。与这样艺术一样的设施相媲美的,是泳池附近的太极园,每天清晨都有太极师傅在此言传身教,试想在85 层的高空迎着日出的光芒领略东方武术,该是怎样一种心情?酒店室内设计由每个城市只做一个项目,纽约最成功的华裔设计师李裕堂(Tony Chi)完成。"他延续了中国式私人住宅理念。"

Park Hyatt Shanghai occupies the floors of 79 to 93 in Shanghai World Financial Center buliding which is also known as the Vertical Integrated Community, as if it is standing on the shoulder of a giant capable of looking over the scenery on both sides of Mitsue. It is defined as a modern chinese-style boutique residential hotel showing its low-key luxury.

The designer cares not only the interior design but also the enjoyment and comfortability of the customers, so a growing lush bamboo forest stand here leading us to the ideal destination in our mind.

The hotel's overall design is maily in gray, white and brown to reach a sense of thrift and low-key luxry and a harmonious and elegant color effect, and to build a chinese ink painting freehand by fusing creative art works in the designing.

Its classic lies in the perfect expression of the modernistic chinese pattern of life without any typical chinese symbol, which subverts the chinese ancient style and presents itself with quite a modernistic view.

If based on the home feeling to re-examine the bottom section of the design is exactly the way cleaning up your mind to home, while every space shared by two rooms is just a small courtyard. The small lawn at the entrance to the suite provides the customers a garden-home feeling, coupled with the Tai Chi area in the public facilitiesthe distribution of the whole trend is like to enter at the gate, pass the hall and rooms, and finally end in the garden.

Because of the height of more than 400 meters, Park Hyatt Shanghai owns the world highest swimming pool. The clear blue above the lights of the houses is just like the sea water floating in the sky of this city. It is undoubtedly a more vivid artistic conception at Park Hyatt Shanghai, that the blue body of water lies still between the white in the room and the black outside, just like a pure work of art carrying a mysterious power straight to the heart. As atrractive as such artistic facilities is the Tai Chi park near the swimming pool. Every morning Tai Chi masters exercise here. Imagine what a pleasant mood it is to enjoy oriental martial arts in the high altitude of 85 layers against the sunrise light. Hotel interior design is done by Tony Chi who is one of the most successful ethnic Chinese designers in New York and only does one project in each city. It continues the idea of China's private residential houses.

餐厅以简单的色调，精炼的图形语言，再加上质感丰富的材料，组成了整个空间温馨而舒适的现代就餐环境。尤其配上巨大的玻璃幕墙，让就餐者一边享受美食，一边俯瞰浦江两岸，惬意而自在。

The restaurant entirely provides a modern and comfortable dining environment constituted by the simple tone, refined graphics and a wealth of material texture. Coupled with a huge glass wall, it makes the customers capable of enjoying the Huangpu River over their food, which is cozy and comfortable.

酒吧的品质在设计中一样举足轻重，并不是应用最昂贵的材料，才能显示它的魅力，而是如何将材料本身的美感发挥到极致才是设计师最终要诠释的目的。它就这样展现在了我们面前。

The quality of the bar is as important in the design. It does not neccesserily depends on the application of the most expensive materials in order to show its charm. What the designer finally wants to interpret is how to make the material to its ultimate beauty, thus it shows itself to us.

空间界面异常的干净、简洁，但并不空洞，并配有前卫艺术家的艺术作品和简洁、独特的灯饰，将他们整合起来就是柏悦表现低调的同时注入了现代的中式气息。

The interface of the space is extremely clean and simple but not empty at all. It is decrated with works of avant-garde artists and simple but unique lighting, totally expressing Park Hyatt's low-key but modernistic Chinese style.

环境的体现是由光线、色彩、质地及人性符号的注入，这些都是能够体现设计师意图及空间效果的重要依据。昏暗的走廊，具有抽象肌理墙面效果，极具风格的灯饰，都能展现其完美的组合。

Environment is reflected by the light, color texture and intake of symbols of human nature, which are all the important basis reflecting the spacial effect and the intention of the designer. The dark corridor assuming abstract wall texture effects and the highly stylized lighting can both demonstrate the perfect combination.

中国元素渗透到设计内部,随着设计的风格变化着,创新着。现代社会适应现代青年的喜好,因此设计满足适用群体,满足社会发展。设计师也就在这一过程创造出越来越多的精美作品。

Chinese elements penetrate into the internal design. changing and innovating as the designing style varies. As modern society adapts to modern young people's preferences, the design should meet the needs of the consuming groups and the needs of social development. Also, in this process the designer creates more and more exquisite works.

客房延续着整体的素雅,白与灰的基本色调,入口处别具匠心的铺设了一小块草坪,犹如进自家的后花园一样,直延棚顶的门与柜,始终维持着一种比例关系,呈现出中国式的住宅理念。

The rooms remains the overall elegant and the basic colors of white and gray. a small lawn is ingenuinly paved at the entrance, as if you are into the back garden of your own. The door and cabinet directly extending to the roof remains appropreate proportion, showing a concept of Chinese-style residential houses.

卫生间和卧室的搭配一致,且具有各种高级设备及用品,除了满足客人使用的需求外,还温馨的提供了各项周到的服务。

The arrangement of the bathroom match with the bedroom, with a variety of advanced equipment and supplies, In addition to meeting the needs of guests, it provides various warm and thoughtful service.

PARKYARD HOTEL SHANGHAI

上海博雅酒店

地址：	开业时间：	设计单位：	面积：
上海市浦东新区碧波路699号	2007年	加拿大GLYPH公司	26000平方米
LOCATION:	**OPENING TIME:**	**DESIGN UNITS:**	**AREA:**
699 Bibo Road, New Pudong Shanghai	2007	The Canadian GLYPH Company	26000m²

PARKYARD HOTEL SHANGHAI

博雅酒店位于上海张江高科技园区碧波路上,四周绿茵环绕,景观极具魅力。酒店同时还拥有十分便捷的交通地理环境,毗临地铁2号线张江站,驱车只需2分钟到达磁悬浮车站,也是距新上海国际展览中心最近的酒店之一。离浦东国际机场仅20分钟车程,而虹桥机场只需40分钟车程。驱车或乘地铁前往陆家嘴金融区,仅需15分钟。酒店楼高6层,共有客房总数300间套,标间面积38平方米。客房设施齐备,有独立自控空调、迷你吧、宽带网络接口、电子保险箱、全天候卫星电视、浴室里装备有吹风机和热带雨林喷淋头、开放式的厨房。其他配套设施包括宴会厅、商务中心、恒温室内游泳池和健身中心;酒店全馆均可免费无线上网。

博雅酒店的一大特色是拥有数百个自然景观房,每个房间都融合现代简约的时尚风格,但也不缺民族特色的东方元素融入其中。

高大的玻璃落地窗虽然阻隔了人们与自然的沟通,但它却以另一种方式让我们更加关注自然的变化过程。在休息区中,坐在简约舒适的沙发上享受着窗外风景带给我们恬静、舒适的感觉,也别有一番情趣。

Parkyard Hotel Shanghai is located Bibo Road, Zhangjiang Hi-Tech Park area of Shanghai. It's surrounded by greenery around. The landscape is prominently fascinating and charmng. The transportation nearby is very convenient, with the subway Line 2 at Zhangjiang Station very close and Maglev station just 2 minutes drive to arrive, It's one of the hotels that are closest to the new Shanghai International Exhibition Center. It's only 20 minutes' drive to Pudong International Airport and 40 minutes' drive to Hongqiao Airport from the hotel. It takes 15 minutes to Lujiazui Financial District by car or subway. The is hotel 6-storeyed with a total number of 300 sets of rooms. The standard room occupies 38 square meters area. All room facilities are available including, independent controlled air conditioner,mini bar, broadband Internet access, electronic safe, all-day available satellite TV, bathroom equipped with hair dryer and tropical rain forest shower and open kitchen. Other facilities include a ballroom, business center heated indoor swimming pool and fitness center. Free wireless Internet access is available in the whole hotel.

one of the charicteristics of Parkyard Hotel Shanghai is that it has hundreds of rooms with a view of a natural landscape Each room is a fused into both the modern and simple fashion style and elements of the east national characteristics.

Although the tall glass windows block the connection between people and nature it provides a chance for people to pay more attention to the process of natural change in another way. In the rest area sitting on the simple but comfortable sofa and enjoying the scenery outside the window gives us feeling of tranquility and comfirtbility, which is a particularly delightful.

三片弧形隔断自然的将休息区与大堂划分出来，吊顶与其遥相呼应，将休息区无形的包裹了起来，让它那份简单、安逸的氛围永远停留在这个空间之中，不被厅堂中人来人往的熙攘所干扰。

3 arcs naturally cut off between the rest area and the lobby, with which the ceiling invisibly wrapping up the rest area echoes, making the simple and cozy atmosphere keeop staying in this space and not to be disturbed by the bustling crowds.

无论是地面矩形的交错搭配,还是顶棚的线条处理,还有两侧对称的沙发及正中的装饰台,都衬托出服务台的简约与庄重。

Whether the rectangular bricks crossly set in the groud or the arrangement of the roof line, the symmetrically placed sofa on both sides, and the decoration desk in the middle, all contrast to the simpleness and solemn of the service desk.

大堂中成对角摆设的沙发,既是提供宾客休息的家具,也是空间中的一个重要装饰品,在映衬上二层的植物及顶棚倾泻而下的灯光,都在述说中国情调。

The sofa diagonally furnished in the lobby, is important both for guests to rest and to be a decortion to the space. The plants of the second floor and the lights pouring down from the ceiling, both tell the Chinese plot of the design.

墙面处理的简单而有变化，吊顶也只限于基本的线条排布，家具精炼、利落的至于空间中心与两边，让整个会议讨论都能在严肃的环境下有节奏、有效率的进行。

The wall treatment is simple but diversed. The ceiling is also limited to the basic line arrangement. The furniture is refinedly and neatly placed in the center and on both sidesof the space, so that the entire meeting is rhythmicly and efficiently conducted in a serious and solemn environment.

自然的光线,葱郁的翠竹,可以让人放松,再加上室内空间的简约设计更加符合人心里的感受,舒适中仿佛回归自然,与外界纷繁熙攘隔绝,尽情的享用就餐的愉快。

Natural light, lush green bamboo, can make people relax. Moreover the simple design of interior space conforms to the feelings of one's heart, as if returning to the nature, isolated from the bustling outside world and just enjoying the dining pleasure.

简单的色调、简单的图样、简单的装饰,给人清新、干净的视觉感受,像是在室外回廊中漫步,远处还有几棵高大的树木在那里静候我们的到来。

Plain colors, simple patterns and decorations provides a fresh and clean visual experience, as if you are walking in an outdoor gallery ,in the distance of which there are some tall trees await your arrival.

客房设计依然传承酒店整体的设计风格,简单中求变化。让宾客在繁忙紧张的工作结束后,能感受到家的舒适与温馨,最终获得身心的释放。

The room design still remains the overall design style of the hotel, a pursuit of diversity in simplenesss. So that the guests can have a comfortable and cozy home feeling and the finally release their mind after the high tension of work.

INTER CONTINENTAL SHANGHAI PUXI

上海浦西洲际酒店

地址： 上海市闸北区恒丰路500号	开业时间： 2010年	设计单位： Lim,Teo+Wilkes Design Works Pte Ltd	面积： 127800平方米
LOCATION: No. 500 Hengfeng Road, Zhabei District, Shanghai	OPENING TIME: 2010	DESIGN UNITS: Lim,Teo+Wilkes Design Works Pte Ltd	AREA: 127800m²

INTER CONTINENTAL SHANGHAI PUX

Intercontinental Hotel Shanghai Puxi In Zhabei District is located in the central business district with superior geographical environment and convenient transportatiion, Shanghai railway station, elevated rail transit and ground transportation network extending in all directions. The hotel has a number of 533 deluxe rooms and suites, including 25 administrative suites, 3 duplex suites, 2 presidential suites and an intercontinental suite. In addition, in the independent administrative building there's a luxurious spacious Intercontinental Club Bar, supporting other facilities. The hotel is very close to Shanghai Railway Station and subway station. It's only 2 kms far from the Shanghai Grand Theatre and the Jing An Temple, and 3 kms from the Bund, 4 miles from the Yu Garden and the Oriental Pearl TV Tower. Moreover. You can walk to the Jade Buddha Temple in 15 minutes and drive to Pudong International Airport in 60 minutes ,or drive to Hongqiao International Airport in 30 minutes.

The exterior wall of the hotel is glass-enclosured. Seen from afar, the two towers of glass and steel structure are the most prominent mark of the hotel and one of the most fantastic landscape around the railway station.

The overall design of the hotel has a blending of postmodern effect with Chinese traditional style and modern minimalist style,which show the demeanor and luxury. At the top of the Central Hall what looks like a Chinese traditional white ribbon flows,which is actually made of wires and glass tubes. The guest in the hotel would feel as if he is a fairy walking in the clouds on the earth.

位于闸北区的上海浦西洲际酒店坐落在市中心商业区，地理位置优越，交通便捷，上海火车站、城市轨道交通和地面高架道路组成的交通网络四通八达。拥有533间豪华客房和套房，其中包括25间行政套房、3间复式套房、1间洲际套房及2间总统套房。此外，独栋的洲际行政楼内设有豪华宽敞的洲际俱乐部酒廊，配套设施一应俱全。酒店毗邻上海火车站及地铁站，距上海大剧院和静安寺仅2公里之遥；距外滩3公里；离豫园和东方明珠电视塔仅4英里。不仅如此，从上海浦西洲际酒店步行15分钟便可到达玉佛寺。至浦东国际机场仅60分钟的车程，或驱车30分钟便可抵达虹桥国际机场。

酒店外墙是以玻璃幕墙结构围合而成，远处看去，两个玻璃和钢结构结合的高塔是酒店最显眼的标记，也是火车站周边最亮丽的风景线之一。

酒店的整体设计采用了中国传统风格与现代简约式风格相融的后现代主义效果，大气中渗着豪华气派。中央大厅顶部飘动着由钢丝结构加玻璃管制成的具有中国传统风格的白色飘

设计师在大堂的布局和配饰上运用了大量的中国元素。高大镂空的金属屏风，丝毫没有给人以压抑感，反而增强了空间的进深感。入口处的假山更体现了中国古典园林的特色。

The designer used a large number of Chinese elements into the layout and accessories of the lobby. The high hollow metal screen by no means gives a sense of oppression, but enhanced the visual depth of the space. The rockery at the entrance also reflects the characteristics of Chinese classical gardens.

酒吧周边运用了大量的大理石彰显酒店的豪华与大气。酒柜更是从地面一直延伸至顶部，让人不得不佩服设计师大胆的创意，也昭示出中国人心胸的宽广与豪爽。

Aroud the bar large amout of marbles are used to highlight the luxury and demeanor of the hotel. Wine cooler extends its way from the groud directly to the roof, pushing you highly respect the bold ingenuity of the designer, and simeoutanieously showing the forthright and broad mind of our chinese people.

青铜人物雕塑无论从质感还是造型都衬托出了老上海丰富的文化底蕴。灯饰上的选择也是独一无二，别具特色。仰望上去像宇宙中的银河，也像空中划过的烟花，美丽而又充满了无限梦幻。

Bronze figure sculptures bring out the diversed culture of old Shanghai in terms of texture and shape. The lighting is unique and distinctive. Seen from lowerposition it's just like the Milky Way flowing in the universe or fireworks shinging in the sky, which are beautiful and full of fantasy.

通透的空间氛围,简约的基调处理,再配以线条简洁的桌椅,使整个环境变得如此悠闲,如此惬意。空气中透露着无比轻松的感觉。

Permeability and simplicity of the space environment, together with neat lines of furniture make the whole environment become so relaxing, comfortable and pleasant. You can feel an unparalleled sense of ease.

简单的几何图形,围绕出丰富的空间环境,这也是设计师的一大妙笔。矩形的吊顶,餐台,配以弧形金属网的餐区,让整个就餐过程变得简单而又私密。

Simple geometric graphics around the large space environment also shows the great talent of the designer. Rectangular ceiling, dining table matched with the food area of curved metal mesh make the whole dinging become simple and self-enjoying.

青花瓷的朴实无华，古典屏风的高贵典雅，都呈现出一种和谐之美，宁静之美。在空间分隔中屏风也起到了重要作用，将过廊与大厅以一种似隔非隔，似断非断的方式区分开来。

The plain blue and white porcelain and the elegant classic screen show the harmonious and tranquil beauty. In the space separating the screen also played an important role, making the corridor and the hall seemingly non-separated.

简单的家具造型配以中国传统色彩与质感的布料；灯具造型似数个灯笼组合而成，却更像一件艺术品；以及具有现代画派的中国画饰品，每个设计都体现出现代与中国古典的完美结合。

Plain furniture style is matched with fabrics in traditional Chinese color and texture. The lighting looking like a combination of several laterns is more of a work of art. The Chinese painting accessery assumes the the characters of modern school painting. All design reflects the combination of Chinese classical and modern.

中国传统花纹的运用已充斥着整个空间。镂空的铁艺隔断；软椅上自由弯曲的线条；地毯纹样的连续性，都在陈述着中国的历史。

The use of Chinese traditional pattern has been filled the entire space. Pierced the iron partition, freely curved lines on the soft chair, the connectivity of the carpet patterns all present Chinese history.

灰色的走廊墙壁，混搭着现代抽象派的装饰画，整体色调中求统一，画在不知不觉中已成为墙体的一部分。走廊尽头是可供人们休息的酒吧空间，低吊顶并没有给人压抑感，反而让我们想偷偷的躲在这个没人经过的狭小空间中小酌一番。

Gray corridor walls decorated with the modern abstract paintings reflects the pursuit of unity in the overall hue. The paintings unwittingly integrate to the wall. The end of the corridor is a bar room for people to rest. The low ceiling of the bar by no means gives a sense of oppression, but offers a desire for people to privately have a shallow drink in this small unfooted space.

BODI BOUTIQUE HOTEL SHANGHAI

上海博地精品酒店

地址：
上海市浦东新区龙阳路1255号

LOCATION:
No.1255 Longyang Road, New Pudong, Shanghai

开业时间：
2009年

OPENING TIME:
2009

设计单位：
Pierre Maciag

DESIGN UNITS:
Pierre Maciag

面积：
8856平方米

AREA:
8856m²

BODI BOUTIQUE HOTEL SHANGHAI

上海博地精品酒店紧邻上海世博园区，同时针对展会客人有免费的商务车接送服务，定点往返于新国际博览中心与酒店之间。酒店楼高7层，外观时尚精致，每一间客房都经过精心的布局，将浓郁的法兰西风格和极富现代感的婉约风范融合得恰到好处，距离上海新国际博览中心仅10分钟车程，酒店提供免费班车往返于上海浦东新国际博览中心及世博园区之间。

上海博地精品酒店给人的第一印象是私人会所般的低调与尊贵。此外，酒店还为您特别准备了各式时尚西餐、点心、名酒雪茄、健身桑拿会所及会议室等配套设施及服务。奢华的居家感受和周到贴心的服务，一切尽在博地精品酒店。

上海博地精品酒店大堂设计具有强烈地视觉冲击力的天花色彩加上天花中间一盏水晶吊灯，增加了空间的浪漫情调与氛围。灰色石材地面间隔流畅的弧形线条交织，中心悬吊圆柱形水晶珠帘闪耀着点点光芒，营造出一个时尚而独具品位的酒店接待空间。在电梯口两侧沿用暗花图案的灰镜，其独特的反射效果使空间生动空灵，造成的艺术效果对比强烈。在楼梯右侧，设计师安排了一个红色植物，美化与活跃了大堂的空间气氛，也与水晶吊顶在色彩形成呼应。

Boutique hotel Shanghai is close to Shanghai World Expo area. The hotel provides free bus service for the show business guests, shuttling between the New International Expo Center and the hotel at fixed time. It's a 7-storeyed building with the appearance fashionable and refined. Each room is carefully laid out, properly fusing the deep French style and very graceful modernistic style. It's just 10 minutes' drive to the Shanghai New International Expo Center from the hotel. The hotel offers Free shuttle between Shanghai New International Expo Center and the Expo Site.

Your first impression to Boutique hotel Shanghai is that it's lowkey and dignity just like a private club. In addition. The hotel specially offers a variety of Western fashionable meals, snacks, wines and cigars, fitness club and sauna as well as meeting rooms and other facilities and services. Luxurious home feeling and thoughtful attentive service, you can just find in the Boutique Hotel Shanghai.

The ceiling color with strong visual impact and the crystal pendant in the center of the ceiling stenghthen the romance of space and atmosphere of the lobby. The curved flowing interwoven lines on the gray stone surface of the foor and the suspending cylindrical crystal beady curtain shining in the center create a unique and tasteful hotel reception room. The unique reflection effect of the gray mirror with floral patterns on both sides of the elevator make the space vivid and ethereal, which produce a strong artistic effect of contrast. On the right side of the staircase the designer arranged a red plant to make the space atmosphere of the lobby delightful and vivid, which also echoes with the crystal ceiling in color.

色彩浪漫而典雅，灯光明亮而温馨，映在铝板上梦幻迷离，映在大理石台面上古朴雅致，如流泉叮咚，似浮云蹁跹。设计师用巧妙的手法，解决了不同材料之间的搭配，无雕琢痕迹，和谐而自然。餐厅用丰富的色彩、艺术的灯光、流苏的线条、雅致的纹饰创造出了一个和谐而统一的空间，让食客在品尝美味的同时，又能体味到悠然的享受。

The color is romantic and elegant. The light is bright and cozy, reflecting on the aluminum board blurry and fantastic, on the marble countertop elegant and tasteful, just like the naturally buzzing water or like clouds flowing light. The designer ingeniously dealt with the matching of the different materials without modifying marks, harmoniously and naturally. The restaurant is a harmonious and unified space with rich colors, artistic lighting, tasselling lines and elegant ornamentation so that the diners can enjoy themselves with the delicious food in this tranquil and leisure small world.

长长的通道，在地毯的陪衬下在静谧中流淌着华丽与舒适。在设计中延伸了绚烂色彩，天花镜面的选择，加以昏暗的灯光效果营造了时尚而神秘的空间氛围，丰富空间的艺术效果，大胆的设计手法使客人在空间里感到轻松和惊喜。

Long channel flows peace, comfort and magnificence in the backdrop of the carpet. Gorgeous colors extends in the design. The use of the ceiling mirror togetherwith the dim lighting create a stylish and mysterious space atmosphere, abundant space for artistic effect. The rich aristic effect and the product of the bold design techniques make guests feel enjuoyable and delightful in the space.

客房具有浓郁的法兰西风格和极富现代感的婉约风范,选用深色木制家具及极少的金属装饰,色彩以深褐色、红、白、银等色调相配合,通过简洁的造型营造出富有生命力和文化内涵的艺术空间。客房色彩轻松明快,构造上大量直线条的运用,使空间具备强烈的现代感与时尚气质。家私及配饰的布置摆放,看似随意,其实都是设计师对比例精确拿捏的体现。

Rooms has both a very strong French style and graceful modernistic style. It's furnished with dark wood and very few metal decorations. The major used colors are dark brown, red, white and silver. The room is created as a artistic space with a simple form full of vitality and cultural context. The colors are bright and delightful. The large number of straight lines uesd on the construction filled the space with strong modernistic and fashionable style. The furniture and accessories seemingly placed at random are in fact the embodiment of the designer's careful consideration for the protportion.

深色的家具、红色调的壁画与地面铺10mm厚红色纹理的地毯使室内整体色彩十分温馨与协调。

The dark furniture, red hue murals and floors covered with 10mm thick red grain carpet make the overall colo warm and coordinated.

大胆采用紫色为主调,局部配以深色作为点缀,让空间层次丰富起来。电视背景墙及窗户简化的线条的造型,加上浪漫的紫色,给人的简洁华贵的风范。

Bold use of purple as the keynote accompanied by the parial embellishment of dark hue enriched layers of the space. TV backdrop and simplified lines of the window plus the romantic purple, gives the sense of simplicity and luxury.

客房的设计采用了流线形体和圆型设计，圆床和圆形造型的玻璃搭配，体现了浪漫、时尚的设计元素，床头背景墙的图案和地毯的图形产生呼应，使整个空间充满灵动的趣味，给人们带来新的寄居方式。

Eeach room is designed round as a streamlined body. The round-shaped bed and glasses reflects the romantic and stylish design elements. The graphic design of the bedside backdrop echoing with the graphic design of the carpet make the whole space full of active and ethereal fun, which offers people a new way of sojourn.

THE LANGHAM YANGTZE SHANGHAI

上海朗廷扬子酒店

地址：
上海市汉口路740号

开业时间：
2009年

设计单位：
Duncan Miller Ullmann

面积：
12000平方米

LOCATION:
No. 740 Hankou Road, Shanghai

OPENING TIME:
2009

DESIGN UNITS:
Duncan Miller Ullmann

AREA:
12000m²

上海朗廷扬子酒店毗邻上海地标之一的人民广场，与南京路步行街更是咫尺之遥，占尽地利之便。

每个名门望族背后都有一段传奇故事，贵为豪华酒店集团典范的朗廷酒店集团自然也不例外。朗廷璀璨不朽的传奇始于1865年，当时伦敦朗廷酒店开业，成为欧洲首家豪华酒店。伦敦朗廷酒店作为朗廷酒店集团的旗舰酒店，曾接待过不少欧洲皇室贵族、达官显贵和文艺才俊，他们无不为酒店触动他们视觉、听觉、嗅觉和味觉的巧妙设计而赞叹。今天，朗廷酒店集团的网络遍及全球，无论在旗下任何一间酒店，独特的瑰丽显赫设计、创新的待客之道和体贴挚诚的服务，也会令客人眼前一亮。

上海朗廷扬子酒店是由曾有"远东第三大饭店"之称的扬子饭店，和拥有140年历史的朗廷酒店集团的结合体。并

请来精品酒店领域的设计权威Duncan&Miller Design，将老建筑重新以新的面目展现于世人。

设计师在设计上继承了英国豪华酒店的传统，并加入了老上海和Art Deco的设计元素，使整个酒店拥有高贵典雅的气质和浪漫奢华的享受。当客人踏进酒店，朗廷独有的姜花香味随即飘然而至，洗涤疲惫的心灵，酒店将幻化成闹市中的世外桃源。

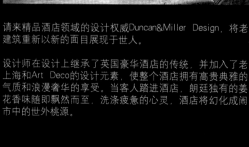

The Langham Yangtze Hotel is extremely well placed adjacent to Shanghai People's Square which is one of the landmarks, and even closer to Nanjing Road Pedestrian.

Each long established family has a legend behind itself. The Langham Hotel as a model for the luxury hotels International is no exception. Langham bright immortal legend began in 1865, when the London Langham hotel opened, which became the first European luxury hotel. London Langham Hotel, as the subordinate hotel of Langham Hotels International, has received a number of European royalties, highbrows and literary talents, all of whom were profoundly touched by the ingenious design of the hotel in all sences. Today, Langham Hotels worldwide networks reach all over the world; Any of its subordinate hotels spark your eyes with its uniquely magnificent design, innovative reception way, sincere hospitality and attentive service.

The Langham Yangtze Hotel is the consolidation of Yangtze Hotel which has been called "the Far East's third-largest hotel," and the Langham Hotels International which has a long history of 140 years. It was designed by the prestigeous boutique hotel designer Duncan & Miller Design, who gave the old building a new look.

The designer followed the traditional design of the British luxury hotel, and added the old Shanghai and Art Deco design elements, so that the whole hotel has its gorgeous elegance and romantic luxury. Entering the hotel you will smell the particular scent of ginger flowers suffusing in the air and cleaning your tired mind. Thus the hotel will be temperarily illusioned as a Xanadu isolated from thhis bustling city.

简单的波浪线型吊顶，整齐、对称的方柱刻画出Art Deco的痕迹，在此之上又赋予繁复、缤纷、华丽的装饰图案，令整个过廊都充满了富贵、华丽的气息。尤其是中国红更显华贵的空间氛围。

The simple linear waving ceiling and symmetrical and neat rectangular columns depicts the traces of Art Deco. Added with the complex magnificent graphic designs the corridor is full of grace and luxury, esp. the China Red presents a luxurious space atmosphere.

沿着朴实中透着华丽的楼梯向上,视觉无限延伸,令我们眼前一亮的是简单的图案和纯净的色彩拼合而成的玻璃天花板,犹如教堂般华丽,顷刻之间就能让自己得到升华。

Tracing up the simple but splendid stairs, unlimitedly extending your vision you will be delighted by the glass ceiling composed of simple patterns and neat colors, just as gorgeous as the cathedral, instantly making yuor soul sublimate.

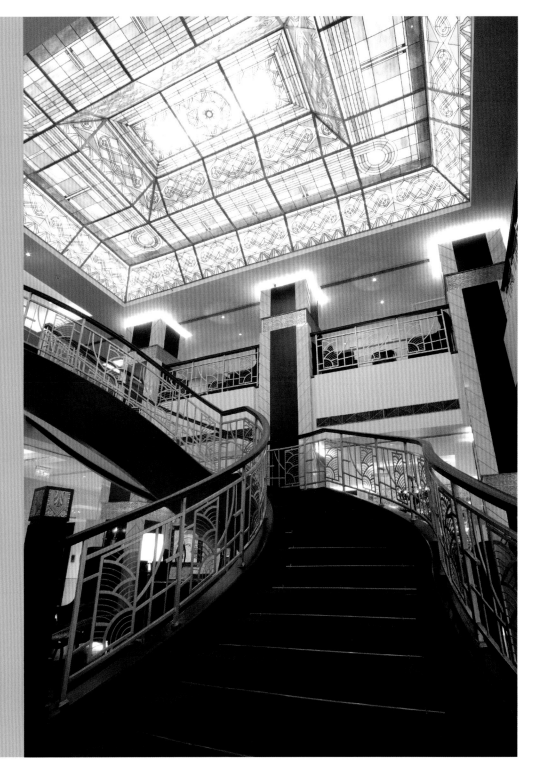

意大利餐厅延续着现代简洁的风格特点，座椅的图案与地面拼贴的花纹有意式的风格，饰品更是能体现出意大利风情，使客人在轻松的环境中享受最正宗的意大利美食。

Italian restaurant continues the simple style of modern characteristics. The pattern designs of the seating and the mosaic floor assume Italian style, and the accesseries further highlight it. The guests just enjoy the most authentic Italian cuisine in such a relaxing place.

餐厅设计一如它的名字——唐阁。设计师在地面、墙面及吊顶中都融入了装饰性很强的中国艺术元素，使整体空间呈现出格调优雅、气派不凡的效果。

Restaurant design just deserves its name - the Tang Cour. The designers integrated into highly decorative elements of Chinese art on the ground, wall and the ceiling, making the overall space presents an elegant and stylish extraordinary artistic effect.

设计师秉承着一贯的装饰手法，简洁的灯具和座椅，以及墙面锯齿状的图案和矩形叠加的隔断效果，再配以深色木材和舒适的灯光，打造出了传统居酒屋的休闲环境。

The designer adhered to the consistent pattern of decorative design approach. The simple and neat lighting and seating as well as the jagged patterns on the walls and the pierced rectangular partition effect added with dark wood and soft lighting create a traditional leisure pub environment.

深色与浅色的结合需要用亮丽色彩来调和,以使整个过廊亮丽起来。对于疗养中心则采用了单一的色调使环境变得更加的宁静,并享受着精神的放松与舒适的享受。

The combination of dark and light needed to be well matched with the bright colors to bright up the whole corridor. As to the Wellness Center the designer used a single color to make the environment feel more quiet so that people can enjoy the spiritual relaxation and comfort.

客房在不脱离整体风格的前提下，呈现出了老上海的瑰丽气息。但设计师从不缺少对细节的考虑，通风口的图案纹样与大厅楼梯的金属栏杆造型就是一个典型的统一。

The rooms show the magnificence of old Shanghai without saperated from the overall design style. But the designers never giving up the considering of the details. The typical uniform of the design patterns of the vents and the modeling of metal railings in the hall is such a good exalmple.

LE ROYAL MERIDIEN SHANGHAI

上海世茂皇家艾美酒店

地址：
上海市黄浦区南京东路789号

开业时间：
2006年

设计单位：
Bilkey Llinas & Associates Robert Bilkey

面积：
55000平方米

LOCATION:
No. 789 Nanjing Road, Huangpu District, Shanghai

OPENING TIME:
2006

DESIGN UNITS:
Bilkey Llinas & Associates Robert Bilkey

AREA:
55000m²

上海世茂皇家艾美酒店作为上海的地标之一，拥有雅致的环境和周到的服务。酒店的整体设计融合了时尚精致元素和超现代设计的理念，使其成为上海最新的白金五星级酒店之一。

坐落于历史悠久的南京东路，正对人民广场，可以俯瞰上海博物馆和上海大剧院。上海世茂皇家艾美酒店周围有高档餐厅、酒吧、商店和博物馆，通过毗邻酒店的南京路步行街或酒店下的上海地铁，您可以畅达上海全城，距外滩只有20分钟的轻松步行路程。上海世茂皇家艾美酒店设施有：洗衣服务、会议室、叫醒服务、理发美容室、停车场、行李存放服务、票务服务、前台贵重物品保险柜、外币兑换、商务中心、残疾人客房、酒吧酒廊、咖啡厅。

酒店定位着眼于年轻和时尚的消费群体，因此设计师在设计时考虑到英国艾美酒店优雅文化的同时，紧跟时代的潮流，注入现代时尚元素，为上海浦西的最高楼赋予了最高的品位与商业生命。

Le Royal Meridien Shanghai as a landmark of Shanghai provides elegant environment and attentive service. The overall design of the hotel combines exquisite fashion elements and ultra-modern design philosophy, making Shanghai one of the latest platinum five-star-rate hotel.

Located in the historic Nanjing Road, facing the People's Square, the hotel overlooks the Shanghai Museum and Shanghai Grand Theatre. Le Royal Meridien Shanghai is surrounded by upscale restaurants, bars, shops and museums. From the hotel you can smoothly reach every part of this city through the Nanjing Road Pedestrian Street adjacent to the hotel, or the subway under it. It's just 20 minutes easy walk from the hotel to the Bund. Le Royal Meridien Shanghai hotel supplies various facilities and services including laundry, meeting rooms, wake-up service, hair beauty salon, parking, luggage storage, ticketing service, valuables safe at the front, currency exchange, business center, disabled room, bar and lounge and coffee shop, etc.

Hotel is positioned to the young and trendy consumers, so the designers not only took into account the elegant British culture of Le Meridien but closely followed the trend and added into the modern fashion elements in the design concept, giving the highest building in Puxi, Shanghai the highest taste and

设计师通过材料的特点及质感塑造出即丰富而又有层次感的墙面效果，给酒廊增添了一种大气与时尚的气息。玻璃踏步与护板组成的旋转楼梯，更衬托出现代的装饰语言。

The designer created diversed and layering wall effects properly making use of the material characteristics and the texture, dignifying the lounge with a stylish atmosphere. The spiral staircase composed of glass steps and protection boards brings out the language of modern decoration.

互相交错的吊顶，与地面的的分区构成了统一的语言。墙面不规则石材的拼贴，与几何图形的地面遥相呼应，共同诉说着自由与闲散的乡间情调。

Interwoven ceiling integrated into the saperated ground discribs a unified language. The wall surface of irregular stone collage and the ground of geometry grphics echoed with each other, dipicting the countryside ambience of freedom and leisure.

设计师在中式餐厅的设计中,中国红的运用恰到好处,且点到为止,既体现了中国元素,又体现出了时代感,时刻提醒着宾客置身在中国最时尚的城市之中。

The use of Chinese red is just right in the Chinese restaurant design, and not more or less than it should be. It not only embodies the Chinese elements, but also demonstrates a sense of the times, reminding you of being in the most trendy city of China at any time.

中式餐厅的入口以不同视角向我们展现了中国水池中嬉戏的金鱼，它将原始的水池变成了玻璃的方形拱门，以现代的手法诠释古典的意境。

The entrance of the Chinese restaurant shows us the goldfish playing in the China's pool from different angles. The designer originally made a pool into a square glass arch, interpreting of classic mood in the modern way.

乡间小路的那份自在与清新，宾客同样可以在高级酒店中感悟的到。酒店的廊道边采用了四道卵石作为镶嵌，好像道路两旁的石子，鲜花在石子中争相开放，让过往的行人不时的驻足观望。

You can just enjoy the cleaness and leisure in this top hotel, which can be only found on the country road. Four lines of pebbles are symmetrically mosaicked on both sides of the corridor. Flowers blooming out of the pebble lines are so atrractive for the people to stay and enjoy.

酒店的大堂独具特色的是服务台上方的方形灯箱，图案取材于芦苇秆子的形状，交错排列，它与客房走廊的天花与地面有着异曲同工之妙。既别致，又富有创意。

The uniqueness of the lobby is the rectangular lightbox above the service desk. The concept of the patterns are drawn from the shape of the reed stalk. They interwave with each other, resulting the same artistic effect with the ceiling and the ground of the corridor, unique and creative.

电梯厅一侧采用透光云石效果，灯光柔而大方且具有十足的现代感，像是能够呼吸一样，让整个空间都不在沉闷。

At the side of the elevator translucent marble effect is used. The lighting soft and luxurious makes you feel modern as if it can breathe and remove the dullness and boredom of the space.

客房的整体呈暖黄色，温馨又舒适。卧室的墙面以淡颜色的木饰面为主，采用了白影及少量的黑檀，让木香味萦绕在床头，令宾客安然入眠。

The room is in the color effct of warm yellow, cozy and comfortable. Bedroom walls are decorated with light colors of wood surfaces. A small amount of White Shadow and ebony is ued, whose fragrance linger around the bed to lead you to the nice dream.

GALLERY SUITES SHANGHAI

上海联艺·凯文公寓

地址：
上海衡山路525号

LOCATION:
Hengshan Road 525, Shanghai

开业时间：
2009年

OPENING TIME:
2009

面积：
3000平方米

AREA:
3000m²

GALLERY SUITES SHANGHAI

上海联艺·凯文公寓位于衡山路，亦为老上海法租界，原为俄国公主Olga Gregorievna Ogneff在上海期间的府邸。这座充满传奇色彩的建筑，经过设计师重新的诠释之后，将时代元素渗入建筑之中，使得高科技与艺术元素以新的面貌描绘出了现代人眼中的老上海风情，让整个设计更增添了摩登与现代的实用感。

ART DECO风格被表现于房间与公共空间内，大量怀旧元素，物件被设计师重新安排。其中怀旧元素与现代元素在整个空间中相互影响，又相互渗透，彼此和谐而完美的融入了时代的河流之中。让整个设计重新拥有了modern&chic的力量。

酒店的目标在于引领艺术潮流，推动当代艺术文化的发展。所以在酒店公众区域能够随处可见当代艺术家的天才之作。

Gallery Suites Shanghai. Kevin apartments is located in Hengshan Road which used to be the old French Concession, the living place of formerly Russian Princess Olga Gregorievna Ogneff. This legendary building is fused into modern element after re-interpreted by the designer. The high-tech and artistic elements depict a new modern look of the old Shanghai rhythm to the people of our time, adding practical sense of modernity and fashionability to the whole design.

ART DECO style is represented in the room and public space, and a lot of nostalgic elements and objects were re-arranged by the designer, interacting and interpenetrating with the modern elements, which penetrate harmoniously and perfectly to the age stream, re-injecting the power of modern & chic to the whole design.

The hotel aims at leading artistic trends and promoting the development of contemporary arts and culture. Therefore, in public areas of the hotel contemporary artists of genius is visible to every people.

大堂酒廊没有富丽堂皇的餐厅,但却提供欧陆早餐供客人品尝,同时酒店中央全天候供应咖啡、茶、软饮和新鲜果汁,让客人随时能够品尝到酒店温馨而又周到的服务。大堂柱子采用水泥、玻璃现代装饰材料,使得大堂整体气氛充满了现代气息,非常适合休憩。

There is not a restaurant that is so magnificent in the lobby Lounge, but it offers a continental or European breakfast for guests to savor. At the same time coffee, tea, soft drinks and fresh fruit juice are provided at the center of the hotel allday, so that you can enjoy at any time the warm and attentive service. The hall columns are made of modern decoration materials of concrete and glass making the overall atmosphere of the lobby full of so modern and ideal for leisure.

酒店中不同的客房采用了不同的现代中式元素,以异求同充斥着整个空间环境,更加彰显酒店的主题风格,让人们置身在老上海时代的空间中,流连忘返。

Different rooms use different modern Chinese elements to build the common points in the diversity in the space environment, thus to highlight the theme style of the hotel so that people will linger in the space and forget their way home.

88 XINTIANDI HOTEL

88新天地酒店

地址:
上海市黄陂南路380号

开业时间:
2002年

面积:
3500平方米

LOCATION:
South Huangpo Road 380, Shanghai

OPENING TIME:
2002

AREA:
3500m²

88 XINTIANDI HOTEL

被时尚品牌店及上海城中最热门的酒吧包围的88新天地酒店,向来是世界各地明星的最爱。设计者以体验者的身份进行设计,除了注重硬件设施和设计品味,优质的软环境深得住店客人喜爱。客人下榻的每个房间,都会萦绕着柔和的"Angel"音乐,和怡人的香薰,房内特别安装的空气净化系统保证了每间客房空气清新,此外,E-smog电磁辐射防护系统有效减少了室内有害辐射。更值得一提的是,所有客人在临睡觉前都会在枕头上发现一张"知识卡",上面简短的智慧寓言,在安心入睡时得到心灵的慰藉。为了迎合商务人士的差旅需要,2002年开业的88新天地近期升级了其主要套房硬件设施,并开设全新的88 SPA馆。

上海88新天地延续了新天地中西合璧、新旧结合的风格和理念。此设计选择具有中式元素家具,再将中式园林的造景加以糅合,使空间上有很浓的中式韵味,墙面装饰多以中国画为主。每个房间都经过精心布局,将中国传统元素和异国情调恰到好处地融合,铺陈在帷幔织物和现代风格演绎的仿古家具之间,营造开阔的空间和富有现代感的婉约风范。

88 Xintiandi Hotel Shanghai ,surrounded by fashion brand stores and the most popular bars in town has always been a favorite palce for the stars from around the world. The designer designed as if he himself was just experiencing in the hotel. In addition to the focus on the hardware and the design taste, the high-quality soft environment also won the favoritism of the guests. Each room is flilled with Angel music and pleasant incense fragrance. Each room is equipped with the air purification system to ensure the clean air everyday, in addition."E-smog"electromagnetic radiation protection system effectively reduces indoor harmful radiation. It is worth mentioning that all the guests will find a knowledge card on the pillow before his sleep, and the short fables of wisdom can ease your mind and lead you to a sound sleep. In order to meet the travel needs of business people, the hotel opened in 2002 upgraded its hardware infrastructure of the suite, and put into use the ne 88 SPA Hall.

88 Xintiandi Shanghai inherited the combination of Chinese and Western, of old and new phylosophy. This design selected the furniture with Chinese element, belended with the Chinese garden landscaping endowed the space with a very strong Chinese flavor. The wall decorations are dominated by Chinese paintings. Each room is carefully arranged perfectly blending the traditional Chinese elements with the exotic style, presented between the modernisticly interprated antique furniture and the curtain fabrics to create an open space and the very contemporary graceful style.

设计师在造型上仍保留着中式风格，精雕细刻。大胆使用大理石作为主要装饰材质，运用不同工艺和手法把简单的材质放大效果，从而产生良好的视觉效果。各种朝代的工艺品，实木的家具，在精心的摆放下，使房里各处都充满了中式气息，精美而简洁的中式门，吸收了古典的神韵，又体现了现代的人文。墙面则使用中式风格的元素造型，整个空间里得到了充分的体现，是自然美和环保的完美结合。红色地毯的使整个空间显出活力。

Designer retains Chinese style and the exquisite carving in the structure. The bold use of marble as the main decorative materials and the use of different techniques to exaggerate the effect of the simple materials, which results in good visual effect. Crafts of various dynasties and good-quality wood furniture carefully placed to make the room full of a deep Chinese flavor. The beautiful and simple Chinese-style door not only absorbs classical charm, but also reflects the modern humanities and culture. The wall is modeled with Chinese style elements which are manifested in the whole space, a perfect combination of natural beauty and environmental protection. Moreover the red carpet show the energy of the entire space.

过道以中式元素的家具，门配有红色地毯，红色与棕色的搭配，华丽中透着温和典雅，如梦如幻的感觉，使人浮想联翩，仿佛时光已经倒流回到远古。

The furniture of Chinese elements in the corridor covered with a red carpet directly to the door, and the well-matched red and brown, all demonstrate the elegance in the luxury, leading you to the fantastic imagination, as if time had gone back to the ancient times.

就餐区提炼中式风格的元素作为柱子,墙面造型,精细雅致。地面铺灰色的地毯,简约的餐桌椅配有红色靠垫以及红色的背景墙,将中式元素与多元化的现代元素加以完美的结合。

The dining area used refined Chinese-style columns. The walls are elaborately and elegantly shaped. The ground is covered with gray carpet. The simple tables and chairs are matched with red cushions and red backdrop. All these design depicts the perfect combination of Chinese elements and diversed modern elements.

走廊沙发、楼梯一侧的背景墙与饰品的细节也充分体现了流畅的线条感,弧形的元素被不断地重复,赋予和谐之念,橙色墙面的"跳入"使整个空间顿时温暖欢跃起来,显现出活力与张扬。

The sofa in the corridor, the bakdrop wall at the side of the stairways as well as the details of the accessories further reflected a sense of smooth lines. The arc element is repeated to dipict the harmony concept. The interjection of the orange wall makes the entire space at once warm and vivid, showing the dynamic and promoting atmosohere.

此设计基于中式风格的提炼和升华,采用现代主义的线条,简约的特点,在简约中加入更多的时尚元素,如绚丽的色彩造型,是现代主义在空间中的新表达。

This design is based on the extraction and sublimation of the Chinese style. It's full of lines of modernism and simplicity in which fashionable elements are added. For example the blending of the brilliant colors is a new way of expression of the modernist in the space.

PANGU 7STAR HOTEL BEIJING

北京盘古七星酒店

地址：
北京市朝阳区北四环中路27号

开业时间：
2008年

设计单位：
Ricardo Bello Dias

面积：
48500平方米

LOCATION:
No.27 of North Fourth Ring Road, Chaoyang District, Beijing

OPENING TIME:
2008

DESIGN UNITS:
Ricardo Bello Dias

AREA:
48500m²

BOUTIQUE HOTEL

PANGU 7STAR HOTEL BEIJING

盘古七星酒店拥有新颖的建筑风格和室内设计，是中国传统精髓与现代豪华风格在21世纪的完美结合。这些设计也使酒店成为除了帝皇宫殿紫禁城外，北京最为尊贵的地方。

盘古七星酒店的室内设计出自巴西籍意大利设计师Ricardo Bello Dias的手笔。Bello Dias来自米兰，1990年毕业于巴西建筑学院（Faculty of Architecture），是意大利著名室内和家具设计师Piero Lissoni的入室弟子。这是他在中国的首个项目，其室内设计在中式典雅魅力中融入西方现代元素，尽显高贵。

整间酒店斥巨资打造，随处可见的大理石全部进口自意大利，由本地雕刻大师以手工雕上中国传统图案。现代的家具摆设、照明系统及先进设备均由全球领先品牌设计，例如灯饰品牌Baga及曾与Bvlgari酒店合作的意大利照明公司Metis。餐厅选用Moroso的豪华装饰，客房则配备美国白宫御用的Baker家具。

北京盘古七星酒店是唯一一间获故宫博物院授权，可复制紫禁城内价值连城的艺术品的酒店。这些出自宋徽宗与恽冰等书画大师之手，来自唐、明、清等朝代的珍品，在意大利以铜板进行复制，现展示在酒店的所有客房和公共空间里。

盘古七星酒店所在的盘古大观，是一座引人注目的龙形建筑。这座综合性的建筑由台湾建筑师李祖原先生设计建造，他曾经打造了世界第二高楼——508米高的台北101大楼。

李祖原先生结合其标志性的中国后现代主义理念，将整座建筑设计成雄伟的五座大厦，五座大厦以全长411米的龙廊相连接，龙廊以66根方型花岗岩石柱支撑。每根石柱顶上均竖立一个三米高的龙形雕刻，这些雕刻均以形成时间超过五百万年的整块花岗岩雕成。盘古大观宛如一条蜿蜒盘旋的华夏巨龙，191米的"龙头"部分是一座45层高的5A级写字楼；"龙身"部分则是豪华公寓、餐饮设施及高级购物中心，而21层的盘古七星酒店当属盘古大观的巅峰之作。

Pangu Seven Star Hotel has initiative architectural interior design, representing the perfect combination of the Chinese traditional essence and modern luxury in the 21st century. The design of the hotel makes it the most distinguished place in Beijing second only to the Forbidden City which is the Imperial Palace.

Pangu Seven Star Hotel was designed by the Italian interior designer Ricardo Bello Dias, who has a Brazil Nationality. Bello Dias was from Milan. He graduated from Faculty of Architecture In Brazail in 1990. He was the apprentice of Piero Lissoni,the famous Italian interior and furniture designer. Park Hyatt Shanghai is his first project in China. He fused modern Western elements into Chinese elegance and charm,mannifesting the grandeur of the design

The entire hotel cost huge sums of money. Marbles visible everywhere were all imported from Italy and carved traditional Chinese designs by a local master of. Modern furniture arrangement , lighting systems and advanced equipment were all designed by the world's leading brand design company, such as lighting brand company Baga, and the Italian lighting company Metis which had been cooperated with Bvlgari Hotel. The restaurant was luxuriously decorated with Moroso. And the rooms are furnished with Baker which is appointed by the White House.

Pangu Seven Star Hotel is the only one hotel authorized by the National Palace Museum to copy the priceless works of art of Forbidden City. The treasurary works of Song Huizong and Yun Bing in Tang, Ming and Qing dynasty reproduced in Italy with copper are mow showing in all the rooms in the rooms and public spaces of the hotel.

Mr Li Zuyuan designed the whole architecture as five magnificent high-rise buildings combining with the landmark post-modernist philosophy.They are connected by a 411-meter-long corridor and supported by 66 grnite columns.On each top of the three-meter-tall column a dragon sculpture made of over-5-million-year granite is erected. The Pangu Plaza is like a huge winding Chinese dragon. The 191-meter dragon head is the 45-storeyed 5A office building. The dragon body covers a luxurious apartment, the dining facilities and the high rate shopping mall.While the 21-storeyed Pangu Seven Star resturant is definitely the culminated amazing design of the eifice.

酒店以温暖的金色调为主，包括米色的意大利孔石以及充满帝皇气派的淡黄龙纹地毯。大堂的紫檀天花板上雕有《红楼梦》的场景，以坚固的花岗岩龙形雕刻装饰，而充满现代气息的壁灯的灵感则来自中国的传统乐器琵琶。

The hotel is wholely in the main hue of warm gold, including the beige Italian Ulva and pale yellow dragon-pattern carpet full of royal temperament. The red sandal wood ceiling in the lobby is carved with the scene of "Dream of Red Mansions". Decorated with the hard granite engraved with dragon patterns, the desin of the wall light is full of modernity, while it's inspired by the traditional Chinese instrument of Pipa.

厨师在开放式厨房中为客人准备众多中式、日式、东南亚和西式特色自助美食。明亮的餐厅中摆设着大型装置艺术品，以及由意籍巴西设计师Ricardo Bello Dias创作的时髦家具。

Chefs prepare meals of various tastes in the open kitchen, such as Chinese food, Japanese food, Southeast Asia and Western buffet. The bright restaurant is furnished with large-scale artistic installationsas well as the stylish furniture designed by Ricardo Bello Dias, the Brazail designer with Italian nationality.

精美的私人包间，传统英式风格的烛台、餐厅主人的众多藏书以及用华丽的天鹅绒制成的扶手椅装点着英式包间，均采用意大利家具和日本鸣海瓷器，配有索尼平板电视和独立洗手间。柔和的奶油色是包间的主调，并有耀眼金箔点缀其中。

Sophisticated private rooms, traditional English-style candle holders, and the books collected by the restaurant owners as well as the decorative armchairs made of gorgeous velvet, are well-matched Italian furniture and porcelain in Japan Narumi. The room is equipped with Sony flat TV and independent bathroom. The soft milk color is the main hue of the room, embellished with dazzling gold.

秉承简约的禅意美学，穿过传统日式暖帘后即可见到餐厅的榻榻米和浅色木材装饰，以及巨大的景观窗户，客人就坐在有地下供暖的日式矮桌旁，享受穿着传统和服的日本女侍应的细致服务。部分包间的隔音纸墙可通过按钮收起，合并成一个更宽敞的用餐空间，岚山包间正对着露天的舞台，客人可以欣赏以盘古大观作舞台背景的日本传统歌舞。

The hotel is adhering to the simplicity of divine aesthetics. Through the traditional Japanese warm curtain you can see the tatami, the light color wood decor, and the huge windows of the landscape. Visitors can seat at the Japanese low table above the underground heating and just enjoy the meticulous service of the waitresses wearing traditional kimono. Part of the paper insulation between the walls can be put away as the button is pushed, to form a more spacious dining room. Arashiyama rooms are faced with the open Noh stage so that the guests can enjoy traditional Japanese dance on the stage against the background of Pangu Plaza.

会议室通过不同造型和色彩展现出了不同的空间风格。造型独特的吊顶及红色尽显中国古典韵味；简约适用的吊顶及深色皮质的转椅彰显西方现代手笔，完美的提供给宾客自我需求的满足。

Meeting rooms show diversed styles of space with virious shapes and colors. The uniquely-shaped ceiling and red color in the room manifested Chinese classical flavor; the simpleness and the practicality of the ceiling and dark color leather chair demonstrates the modern Western design concept,perfectly meeting the self-satisfaction needs of guests.

大型报告厅整体采用了红色系为主的空间色彩，集聚象征性的诠释了中国传统色彩的意喻。独特的吊顶造型与简约图形的地面；古典画作与现代"水立方"建筑，都突出了古典与现代共存，时间与空间交融。

The large lecture hall is based on a red color line, symbolicly interpreting the soul of traditional Chinese color. Unique ceiling design contrasting to the simple graphics of the ground; the classical paintings contrasting to the modern "Water Cube" architecture, etc, all demonstrate the coexistence of the classical and the modern as well as the compatibility of time and space.

欧洲进口的黄色大理石被大量运用在走廊的墙面与地面，使酒店的豪华与精美尽情的展现在宾客面前；墙壁的壁灯和顶棚的吊灯，都以中国元素为模板加以西方现代的装饰手法共同谱写出中西合璧的完美。

The walls and the ground in the corridor largely using European imported yellow marbles fully show luxury and exquisition of the hotel to the guests. The wall lights and the pendant of the ceiling use Chinese element as a template, penetrating the Western modern decoration, to depict a perfect combination.

客房采用带圆形凹槽的天花板,象征"天圆地方",籍此增强房间的风水;墙面的古典画作则与灯饰上的龙形图案遥相呼应;欧式家具更以点睛之笔融入整个空间。宽敞的浴室配有豪华的洗浴设备,尤其浴巾架能为精美的Frette浴巾提供加热,让富丽堂皇的客房增添了家的温馨。

The ceiling of the room is round-shaped defining a Chinese traditional concept of "round sky and square ground" hopefully to enhance the geomantic omen of the room. The classical paintings on the wall echo with the dragon design of the lighting. European furniture finally integrate into the entire space. The spacious bathrooms installed with luxurious bath equipment, esp. the bath towel rack to heat the fine Frette towels, add the warmth of home to the splendid rooms.

客房到处都渗透出华贵的气息。温暖的金黄色；古典的龙凤图案；古代著名作家的画作与诗篇，都在展示帝王宫廷的地位与气势。闲散的客厅，高贵的餐厅，舒适的卧室及宽大的浴室，则编织出了一场超豪华盛宴。

Luxury penetrates into everypart of the rooms. Warm golden yellow; classical dragon and phoe all show the position and momentum of imperial court. The leisure living room, elegant resturant, the comfortable bedrooms and spacious bathrooms constitue a super-luxurious banquet.

HILTON WANGFUJING BEIJING

北京王府井希尔顿酒店

地址：
北京东城区王府井东街8号
LOCATION:
No.8 in Wangfujing east street, east area of Beijing

开业时间：
2008年
OPENING TIME:
2008

设计单位：
威尔逊室内建筑设计公司
DESIGN UNITS:
Wilson Interior Design Firm

面积：
38193平方米
AREA:
38193m²

HILTON WANGFUJING BEIJING

北京王府井希尔顿酒店选址于北京最为闻名遐迩的繁华商业区王府井大街，周边著名商厦林立，步行可达紫禁城、天安门广场等中国特色景点，占尽城市商业及文化中心的绝佳位置。

北京王府井希尔顿酒店内部设计时尚别致，其灵感源自经典的洛可可风格以及毗邻的紫禁城古建筑群。北京王府井希尔顿酒店大堂和前台处设计彰显"宾至如归"的主题。

北京王府井希尔顿酒店的整体设计色彩华美且风格多样的画作，由闻名华夏的北京艺术家倾力绘制，也是地地道道的京味儿，同时也令酒店蓬荜生辉。现代风格的深色调实木家具，配以柔和的米色、象牙色织物，尽显舒适之感。酒店内设有四家风格各异的餐厅和酒廊，口味各异的特色餐厅风格时尚别致，典雅且不乏时尚品味，设计注重细节，并在细节处洋溢着浓厚的艺术气息。每一个独具风格和特色的餐厅，配有美味可口的传统及异国菜肴，感受中华传统文化与国际大都市交相辉映的迷人魅力。

Hilton Beijing Wangfujing is located in most famous shopping district of Beijing, Wangfujing shopping district, which is the best location of business and cultural center of the capital city. It's surrounded by well-known commercial buildings. From the hotel you can just easily walk to the Forbidden City, Tiananmen Square and other attractions.

The interior design of the hotel is novel and special, inspired by the classic Rococo style and the ancient Forbidden City architecture. The design of the lobby and the reception area emphasizes the "home" theme.

The paintings of Hilton Beijing Wangfujing with gorgeous colors and a variety of styles are effortully worked by Chinese renowned artists in Beijing, thus assumes a downright flavor of Beijing and throws luster to the space. Modern style furniture in dark colors accompanied by soft beige and ivory fabric provides a full sense of comfort. There are four restaurants and lounges of different styles in the hotel. The novel and chic restaurants provide food of varying tastes, elegant and without lack of fashion sense. The design pays more attention to details and filled them with heavy aristic taste. In the restaurants of various styles and characters that provide delicious traditional and exotic dishes, you will enjoy the charm of the traditional Chinese culture of this metrapolitan.

温馨的壁炉让人们顿生融融暖意，烘托出酒店热情欢迎四海宾朋的待客之道，现代风格的真皮家具，搭配蜜色、象牙色及巧克力色调，讲求整体观感效果，尽显舒适自然。

The fireplace mamkes you feel warm and cozy, displaying the hospitality of the hotel to welcom the universal guests and friends. Modern style leather furniture matched with honey, ivory and chocolate colored tone highlights the overall visual effct and makes you feel natural and comfortable.

北京王府井希尔顿酒店的电梯间不同于通常酒店将电梯间设计得比较昏暗，北京王府井希尔顿酒店的电梯间走廊很明亮，而且还有一个很特别的墙面装饰——凤凰，一个巨大的写意凤凰造型足足占满了一整面墙的空间，图腾式的元素构成了整个凤凰的造型，虽然不是具象，却因到位的神韵使人第一眼便可认出这在中国传统文化中象征吉祥的神物，成了酒店标志性的设计。

电梯口对着的墙面用凸凹不平的造型、线型天花的造型搭配理石地面使空间具有流畅、开阔、柔美的意境，设计用材料的质感与灯光搭配表达出强烈活泼的气氛，整体设计坦率自然又跌宕起伏。

The lift room of Hilton Beijing Wangfujing is more brighter than many other hotels. Moreover there is very special wall decoration - Phoenix. The great freehand Phoenix occupies the entire wall of the room. The totem-like element forms the shape of the Phoenix. It's not so large though, people can recognize the traditional Chinese cultural auspicious religious artifact at first sight, which became a landmark design of the hotel because of its real-like charm.

The entrance of the lift is just in front of the uneven-shaped wall. Linear- shaped ceiling matched with a marble floor endows the room with a smooth, wide and tender atmosphere. The textured materials embellished with the lighting demonstrate a strong lively atmosphere, making the whole design natural, free and waving.

走进餐厅便立刻能够感受到迎面而至的中国元素。先是一把高背儿的木椅，简约古朴，仿佛穿越时空而来；在一张台案上，一盏普通的大红宫灯因为加入了衣衫形状的木质框架而格外吸引眼球，浑然天成的设计使得二者相得益彰；再往里走可以看到一件让人不得不驻足惊叹的艺术品———件两人高的古代官服在餐厅前厅展出，最为特别的是这件巨大的官服是用成千上万枚铜钱组合而成，异常雍容华贵。

You can immediately feel the Chinese elements coming into the restaurant. The first jumps into your eye is a high-backed wooden chair, simple and primitive as if coming through time and space far away. On the table an ordinary red lantern particularly attracts your eyes due to the additional clothes-shaped wooden frame, the totally natural design allowing the two complement each other. Further into the corridor you will be forced to stop by and marvel at the artifact of a two people high ancient official uniform exhibited in front of the resturant, which is specially made of tens of thousands of gold coins, abnormally dignified and elegant.

此设计是中式元素结合以洛可可风格为创作基调,运用现代风格的手法进行重新演绎,适当融入洛可可风格的设计元素,将整体空间中式韵味统一的同时,又让洛可可风格在其中和谐并存,相互融合,交相辉映。

This design is the combination of Chinese elements and the tone of Rococo style, re-interpreted in a modern way. The proper use of Rococo style makes itself coordinated with the unified Chinese elements in space, absorbing and complementing each other.

包容古今，融汇中外文化，在变化中寻求两种文化的结合点并努力贯彻到内部空间每个区域的装饰设计中，力求在追求酒店整体统一风格中体现多元化、多视有的文化内涵，在构造高雅、华贵的酒店的室内空间的同时，创造出别具一格的"现代化传统文化观。"

The design is a good combination of classic and modern, Chinese and foreign cultures to seek the juncture of the two cultures and penetrate them into every detail of the diversifying interior design. It strives to reach a diversified and multi-perspective cultural concept in the unified interior design style. In addition to constructing an elegant and luxurious interior space of the hotel it also aims to create a unique "modern view of traditional culture."

此设计用厚重的实木台板外包大理石,精致艺术的钢铁造型的酒柜,加上晶莹剔透的玻璃瓶搭配背景墙,营造出时尚、舒适、雅致的休息空间。高耸的酒架,质朴的木台,悬挂的装饰品,高低错落的节奏,丰富了空间的层次,华美低调的空间气氛,配上魅力独具的东方鸡尾酒,也许精神上的升华只在这一瞬间凝结……

This design created a stylish, comfortable and elegant space to rest, with a thick wood board wrapped with marble, steel shaped wine shelf of fine art added with crystal clear glass bottles against the backdrop wall. Towering wine shelf, rustic wooden tables and the hanging accesseries all distributed up and down in good proportion, enrich the layering of the space. The gorgeous low-key atmosphere of the space, coupled with the unique charm of oriental cocktail, perhaps only in this moment the sublimation of the spirit condensed....

天花，地面及百合窗以及光色与线条的变换，将空间各层次线与面的交错关系在简与繁之间冲击着视觉，使整体空间在意境中升华。金色的灯光，现代的装饰及线型的设计手法，打破了原先的死板，使整个空间灵动起来，让人的思维也变得活跃起来，餐厅显得雍容而不张扬，色彩层次分明，显得神秘高贵、细致生动而意味深远。

The ceilings, floors and, the screen of the windows as well as the transformation of the light and color lines all impact on the visual effect betweeen the simple and the complex , making the overall space sublime in its atmosphere. Golden light, modern decoration and the linear approach design, enlivens the previously rigid style and makes the whole space vivid, therefore people's thinking becomes active. The restaurant seems graceful rather than gauzy with the color obviously layered. It has a far-reaching meaning tha makes it appears mysterious and noble, vivid and sophisticated.

天花采用巨大的圆形造型与对应圆形造型的餐桌形成呼应，餐厅地面艺术地毯纹样，窗帘的纹样与墙面的图案形成呼应，在色彩上用灯具与装饰品形成呼应，貌似设计随意，简洁大方的设计，其实每一个细节设计师都是经过精心处理的。

The huge round-shaped ceiling echoes with the round-shaped dinner table. The artistic patterns of the carpet and the curtains echoes with the design patterns of the wall which corresponds thelighting and accessories in color. The design seemingly casual is actually simple and elegant, with every detail carefully handled by the designer.

整个设计风格延续酒店完整的造型语言及中式风格，结合现代科技感的手法，利用色彩、灯光、造型把此空间做得亮丽动人。加上玻璃、不锈钢、灯光等现代质感的材质运用，都赋予了它时尚鬼魅、独特新颖的灵魂空间，动感前卫的设计，活力四射的演出，无疑是城中又一"超星级"时尚领地。

The continuation of the complete modeling language and the Chinese style of the design combined with the modern technical design approach, making good use of color and light delights the whole space and makes it attractive. The practical use of the glass, the stainless steel and the light full of sense of texture endows the space with fashion and charm. The unique and creative soul space, the dynamic and avant-garde design and the active performance are undoubtedly another "super star rate" trendy territory of the city.

用纯净、淡雅、明快的米黄色调,衬托高品位的家具、灯具及艺术品陈设的结合突出的了立体感与节奏感,烘托出酒店高雅的文化氛围。

The clear and bright light beige set off the tasteful furniture. The combination of lighting and the outstanding works of art displays the three-dimensional and rhythmical feeling, bringing out the elegant hotel cultur of the hotel.

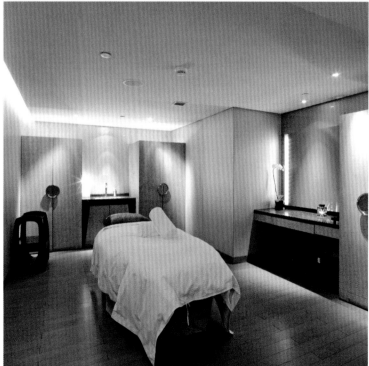

走廊简洁流畅的设计风格,通过艺术地毯的图案纹理贯穿整个空间,同时形成客人的视觉导向,给人们留下独特的印象。走廊材质及色彩的运用上,使用了米黄大理石与米黄色木饰面作为色调配有现代感的地毯,这些视觉感觉温和的色彩及质地较为柔和的材料,更好地阐述了空间概念中"中式儒雅"的主题。

The simple and smooth design of the corridor filling the whole space through the art patterns of the carpet and forming the guiding visual effect leaves a unique impression to the guests. The use of beige marble and timber in the corridor matching with the carpet of modern sense provides the visual sense of tenderness, explaining the theme of "Chinese elegance" in the space.

北京王府井希尔顿酒店的客房风格鲜明且功能齐全。舒适宽敞的客房采用开放式布局，客房面积则为京城之最，最小面积亦有50平方米。灰色调的床搭配中式纹样的灰色地毯，韵味深远。

The rooms in Hilton Beijing Wangfujing hotel is functional and distinctive in style. The comfortable and spacious rooms are openly laid out with the size of each the largest in the capital city. The minimal one occupies 50 square meters. Gray color bed matched with Chinese-style patterns of gray carpet has a far-reaching sense.

北京王府井希尔顿酒店客房的落地窗，自然光线充足，可以一览北京城的美丽风景色。客房内均配备时尚且现代的固件设施和深色实木家具，所有客房均配备有瑰珀翠（Crabtree & Evelyn）沐浴套装、大理石梳妆台、独立双面按摩喷头、漂雨淋浴喷头、全玻璃环绕盥洗设施、维多利亚式迪素（deep-soak）浴缸以及宽敞的衣帽间。

Through the french windows with ample natural light you can fully enjoy the beautiful scenery of the city. Rooms are all equipped with modern facilities and firmwares and dark wood furniture, and provide with Crabtree & Evelyn bathing suits, marble vanities, independent double-sided shower nozzles, all-glass surrounded toilet facilities, Victorian deep-soak baths and spacious cloakrooms.

整体为温暖的现代风格,呼应了整个酒店的设计特点,浅色的椅子,墙面与深色的灯具、桌子、窗帘搭配,配合装饰画和局部地毯,让房间温暖亲切,酒店给人家一样的温暖感觉,让人产生归属感和依赖感。

The overall warm modern style corresponds to the design features of the hotel. The light-colored chairs, walls contrsting to the dark lighting, tables and curtains, matched with decorative paintings and local carpets makes the warm and friendly. So that the hotel gives you a warm home feeling and a sense of belonging and reliability.

THE PENINSULA SHANGHAI

上海半岛酒店

地址：
上海市外滩中山东一路32号

开业时间：
2009年

设计单位：
Pierre-Yves Rochon

面积：
80000平方米

LOCATION:
No. 32 of Zhongshan Dong Yi Road Shanghai Bund

OPENING TIME:
2009

DESIGN UNITS:
Pierre-Yves Rochon

AREA:
80000m²

酒店大楼具有上世纪二三十年代华贵府第的显赫气派，散发浓厚装饰艺术韵味，也是六十多年来首座建于外滩的新建筑。酒店的地理位置非常优越，可尽览历史悠久的外滩、黄浦江、前英国领事馆花园、苏州河以及彼岸浦东的迷人景观。建筑师David Beer和来自PYR S．A．S的巴黎室内设计师Pierre-Yves Rochon，以来自艺术装饰风格的灵感，运用相同材料与设计主题，在上海半岛酒店的主题设计当中，进行了一次简约的现代演绎。

各食府与酒店大楼的设计装潢以上世纪二三十年代贾华宅为蓝本，散发浓厚装饰艺术韵味，重现当年"东方巴黎"的迷人风采。酒店巧妙地将装饰艺术和中国文化元素融入客房的设计之中，共有235间客房，其中包括44间套间，面积傲视优雅尊贵氛围自然流露，却又不失私家居停的亲切贴心。酒店内以礼查、大华、汇中及半岛为名的四间主题套间宽敞豪华，甚有高雅居停风味；身处酒店楼顶露天阳台，江畔外滩、对岸浦东现代化高楼及前英国领事馆花园景致尽入眼帘。

半岛酒店大堂茶座闻名中外，上海半岛酒店自无例外：三层高楼底空间宽敞，光线充裕，全天由早上6时至晚上11时供应各式美食；客人一边轻啜香茶品味下午茶点，或投入传统下午茶舞情调，尽带二十年代优雅风华。大堂旁边的引航酒吧以深紫色及法式航海装潢为题，并陈列了大量海事主题珍贵藏品，气氛舒适亲切，良朋共聚乐无穷。

The Peninsula Shanghai is the ninth luxury hotel under Peninsula Hotel Group.It's also the crystalizaiton of the integrattion of the glorious tradition as well as the luxurious and comfortable accommodation and the advanced technology, and symultaneously the transformation from the former to the latter. It can be respected as the highest hotel level of service In Shanghai.

The hotel building tremendously assumes the prominent luxurious style of 1920s-1930s carrying a strong Art Deco flavor.It's also the first new construction that has been built in the Bund since more than 60 years ago. The hotel location is extremely superior to give a full view of the historic Bund, the Huangpu River, the former British Consulate Garden, Suzhou Creek as well as the other side of the charming landscape of Pudong,. the architects David Beer and the Parisian interior designer Pierre-Yves Rochon from PYR SAS conducted a modern interpretation of simplicity in the theme design of the Peninsula Hotel Shanghai with the inspiration from the Art Deco style and the same materials and design theme.

The interior designs and decorations of the big restaurants and hotels are modeled on the buildings and houses of Chinese merchants of 1920s-1930s,carrying a strong Art Deco flavor and reproducing the charming elegance of the old "Paris of the East," The designer skillfully fused decoration art and cultural element into the design of the guest rooms, with a total number of 235 including 44 suites with grand area naturally disdainful and elegant but without losing the intimacy and cozy feeling of private home.The hotel contains four spacious and luxurious theme suites,viz. the Kalee, the Majestic, the Palace and the Peninsula, providing extremely elegant sojourns. Living on the top of the buliding with the roof terrace,you can have a full glance of the river bund as well as the skyscrapers and the former British Consulate Gardens on the other side in Pudong.

The tea cafe of Peninsula Hotel is world-famous with the Peninsula Shanghai no exception. The first floor of the three-storeyed building is spacious with a high ceiling taking ample light, The hotel serves various types of food from 6:00 am to 11:00 pm , so that the guests can taste the afternoon tea cakes while sipping fragrant tea , or they will be immersed in traditional afternoon tea dance mood, fully reproducing the 1920s' elegance. The piloting bar next to the lobby, with deep purple the key color and French nautical decoration the theme, displays collection of valuable maritime objects to create delightful atmosphere for friends and cronies to have indefinite fun.

坐拥名震中外的上海闹市景致,中式红漆陈设、乌木墙面、光亮的木料及丝绒沙发等装潢同样叫人惊艳,古董及当代陶瓷摆设天衣无缝,与半岛兼融古今的传统呼应。

Sitting among the world-renowned scenes in Shanghai the hotel is decorated with Chinese lacquer furnishings and ebony-covered walls. The polished wood and velvet sofas and other decorations are stunning. The antiques and contemporary pottery are coordinately displayed to echo the traditional blending of the classic and the modern of the Peninsula Hotel.

酒廊洋溢爵士乐韵,驻场乐队及唱片骑师倾力娱宾,轻歌妙舞气氛醉人;客人可以一边轻尝招牌玲珑玫瑰香槟、鸡尾酒或紫色及粉红克巧力,上世纪三十年代纸醉金迷的气氛油然而生。

Jazz rhythm fills in the lounge. The entertaining resident band and DJs provide beautiful songs and elegant dance, so that the guests can taste the exquisite rose champagne, cocktails, or purple and pink Keqiaoli in an intoxicating atmosphere spontaneously sensing the indulging debauchery lif of 1930s.

空间在色彩、材料和灯光的搭配上使整个空间充满亲近、和谐、简洁、现代的感觉，同时还能给人高贵、典雅、庄重的感觉，能体现出公司的形象。再配以简单的条形黑色石材，同时与天花遥相呼应，风格统一简洁而明快。

The proper use of the light, color and materials endowed the entire space with simple harmony, intimacy and modernity without losing grandure to reveal a noble and elegant mood, which perfectly reflects the character of the company. Moreover the simple strips of black stone echoing with the ceiling unified the whole style, concise but vivid.

套间为上海半岛酒店的前身,私人露台配备了按摩浴缸,二者皆能一览黄浦江和浦东闹市的优美景色,宽敞舒适的套间,面积达71—89平方米可饱览外滩以至浦东闹市全景。卧室和客厅布置精巧典雅,而不失优越的空间感。

The suites are the predecessor of the Peninsula Hotel Shanghai, The private terrace is equipped with a Jacuzzi. With spacious and comfortable area of 71 - 89 square meters the suite can have a panoramic view of the Bund as well as the downtown area of Pudong. The layout of the bedroom and living room is sophisticated and elegant, without losing a sense of superiority.

独立书房、客用化妆室、私人健身设备和设备完善的备餐间等,主人卧室更设有特大大理石浴室连按摩浴缸,并引入充沛的自然光线,让宾客尽情投入无比豪华舒适的私人天地。

The independent studyroom, the dressing room, the private fitness equipment and the well-equipped pantry, etc. as well as the grand marble bathroom with a Jacuzzi in the bedroom which takes abundant natural light, all lead the guests into an unparalleled luxurious and comfortable private world.

两层楼底高半岛套间客厅,呈现一派典雅的氛围。而拥有时尚的双层楼高客厅和阳台,把历史悠久的外滩,浦江和浦东美景尽收眼底。

Two-storey-high ceiling of the living room in the suite shows a classic and elegant atmosphere. The stylish double-storey living room and balcony proivde a panoramic view of the historic Bund, the Huangpu River and the Pudong district.

KINGKEY PALACE HOTEL SHENZHEN

深圳大梅沙京基海湾大酒店

地址：
深圳市盐田区盐葵路90号

LOCATION:
No.90, Yan Kui Rood, Yantian District, Shenzhen

开业时间：
2009年

OPENING TIME:
2009

设计单位：
香港郑中设计事务所

DESIGN UNITS:
Cheng Chung Design(HK)LTD

面积：
40000平方米

AREA:
40000m²

KINGKEY PALACE HOTEL SHENZHEN

深圳大梅沙京基海湾大酒店坐落于深圳著名的东部旅游度假胜地——大梅沙海滨，临近大型综合旅游景区东部华侨城、大梅沙海滨公园、大梅沙游艇会、海洋公园、明思克航母世界等。酒店两座主楼体面朝碧波荡漾的大梅沙海景，沐浴在金色阳光、白色沙滩和葱郁园林的包围中，宛如太阳下的一颗粉色珍珠般柔和美丽。酒店283间客房及总统套房坐拥奢华海景和葱郁园林，房间装修典雅时尚，糅合现代风格的超大可开放式卫生间带来耳目一新的感受，每间房的床垫均为特别订制，柔软舒适超越甜梦感受。

Kingkey Palace Hotel Shenzhen is located in the eastern part of Shenzhen, a famous tourist resort - Dameisha beach, close to large tourist attractions of OCT East, Dameisha Beach Park, Meisha Yacht Club, Ocean Park, Minsk Carrier World etc. Two main buildings of the hotel facing the scenery of the rippling blue Dameisha sea is bathed in golden sunshine and surrounded by the white beaches and lush gardens, as soft and beautiful as a pink pearl shining in the sun. The 283 rooms and suites sitting on luxury sea view and lush gardens were decorated elegant and fashionable, and combined with the large modern style bathroom available to the public, it brings a brand new experience. The mattress of each room is custom made, soft and comfortable, bringing experience beyond the Sweet Dream.

日本餐厅的整体设计简洁大方，木质的隔断搭配规则的灯饰，尽显日式餐厅的特殊环境和氛围。这里提供正宗传统的和食美味，此外设铁板烧，临窗而坐，惬意舒畅。

The overall design of the Japanese restaurant is simple and elegant. The wooden partitions matched with regular-shaped lightings fully display its distinctive space environment and atmosphere. Here offers authentic and traditional delicious Japanese food, in addition, it sets Teppanyaki which can be enjoyed by the window, making you feelcozy comfortable.

酒店设有2间风格迥异的酒吧。从早晨到午夜，您都可以在舒适惬意的大堂吧一边品尝各类饮品和特色点心，一边观赏落地窗外的旖旎风景。入夜时分，位于二楼的清吧将为您展现不一样的静谧空间！

The hotel has two bars of different styles. From morning to midnight, you can enjoy the charming scenery outside the windows at the lobby bar over various types of drinks and snacks, which is comfortable and cozy. At night, the pub on the second floor presents you an unusually quiet and private space.

所有客房都将宽阔的海滩视野、个性化的装饰、完美的细节设计、豪华舒适的浴室完美的结合在一起。
Each room is an integration of wide beach view, personalized decoration, perfect details as well as the luxurious bathroom.

客房在颜色与材质的选择和搭配上,整体烘托舒适和沉稳且不失优雅高贵的氛围,灯饰与家具的完美结合,让入住者感受到温馨惬意的感受。

The overall design of color and material selection brings comfortable and calm atmosphere with elegance and nobility. The perfect combination of lighting and furniture brings warmness and coziness to the guests.